World of science

WATER, RIVERS AND OCEANS

BAY BOOKS LONDON & SYDNEY

1980 Published by Bay Books
157–167 Bayswater Road, Rushcutters
Bay NSW 2011 Australia
© 1980 Bay Books
National Library of Australia
Card Number and ISBN 0 85835 274 5
Design: Sackville Design Group
Printed by Tien Wah Press, Singapore.

WATER

All life on earth depends on water. Two-thirds of your body and about nine-tenths of your blood consists of water. Trees and plants need water to grow, and water helps to shape the earth and its islands and continents with their hills and valleys. About 71 per cent — nearly three-quarters — of the earth is covered in water by the oceans, lakes, seas and rivers.

Water and life

Since most living things are largely made up of water, it is obvious that it is very important to life. Humans can live only for seven to ten days without it. In a year, you take in about a tonne of water either by drinking it or by eating foods containing it. A modern family home uses many thousands of litres of water every year.

About half of the material in trees and three-quarters of

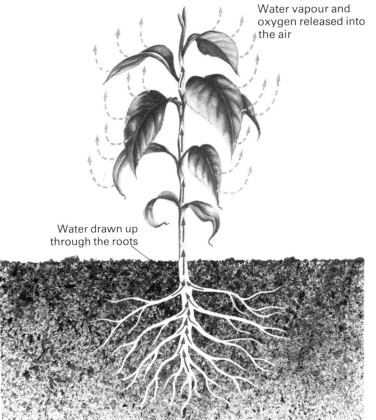

Water vapour and oxygen released into the air

Water drawn up through the roots

Plants depend on the water that they draw up from the soil through their roots. Green plants can make food from this water and the carbon dioxide in the air. This process, whereby green plants convert the energy of sunlight into chemical energy, is known as photosynthesis. Carbon dioxide is taken into the leaf, and oxygen and water vapour are released later into the air.

A rushing river, bubbling over rocky boulders, can gradually weather the rocks themselves. The continuous force of the water wears down the rock surfaces. Exposure to the atmosphere will weather the rocks further as they are gradually broken up or decomposed by rain, ice and natural processes.

Opposite: This enormous glacier is in the far north of Canada inside the Arctic Circle. About one-tenth of the earth's surface is covered by glaciers and ice sheets. Moving glacier ice can actually weather rocks, in time transporting and depositing them elsewhere. Thus they have shaped the landscape of the earth. Many parts of northern Europe, North America and the Soviet Union were glaciated during the Ice Age.

the material in smaller plants is water. Plants have tiny root hairs with which they absorb water containing minerals from the soil. Scientists have found that around 1,000 kg of water is needed to produce one kilogram of plant food.

In various forms, water helps to shape the landscape around us. Water helps to weather rocks, wearing them down and, by freezing in cracks, helping them split apart. Running water in rivers and streams wears away deep valleys and deposits silt, or *alluvium*, on deltas and flood plains. Underground water dissolves minerals and carries them along with it. Ice sheets and glaciers smooth and wear down mountains and deposit silt, sand and gravel elsewhere. The sea wears away some of the coast and dumps the sand and pebbles elsewhere.

The properties of water

Pure water is a colourless, tasteless liquid. It is a compound made up of two atoms of hydrogen (H) with one atom of oxygen (O) so the chemical formula of water is H_2O.

By using electricity, chemists can break water down into the two gases, hydrogen and oxygen, in a process

called electrolysis. Water can be made by mixing hydrogen with half of its volume of oxygen and igniting the two gases with a spark. The explosion produces a vapour which condenses to produce water droplets.

Because many things can be dissolved in water, it is one of the best *solvents.* People who live in areas where bicarbonate of calcium or magnesium from the ground becomes dissolved in water find that soap does not lather well in it. Water containing these minerals is called hard water.

The three *physical states* of water are: water vapour, a *gas;* water, a *liquid;* and ice, a *solid.*

When water evaporates it turns into water vapour and escapes into the air as an invisible gas. Because water absorbs heat when it is changed into vapour, it is possible to cool substances by allowing water to evaporate from them. This is what happens when we perspire on a hot day. Our perspiration evaporates and we are kept cool. Heat speeds up evaporation because the molecules move faster in warm water. The escaping molecules exert a pressure on the atmosphere called *vapour pressure.*

The pressure of the atmosphere affects the tempera-

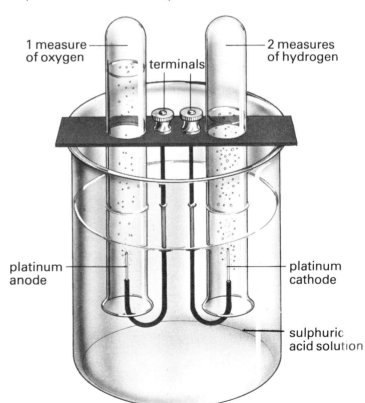

This simple experiment can demonstrate that, by the process known as electrolysis, water can be split into its two constituent gases: oxygen and hydrogen. For every measure of oxygen collected, there are two measures of hydrogen, thus establishing the chemical formula for water: H_2O. There are two hydrogen atoms for every oxygen atom present.

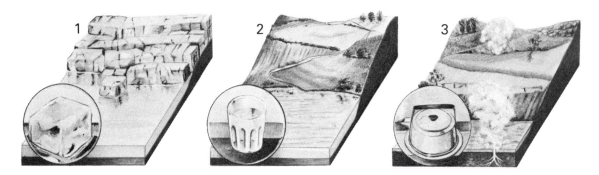

ture at which water will boil. Water boils at the point at which the vapour pressure of the water is the same as the **atmospheric pressure. At sea level, the average atmospheric pressure is 1013 millibars.** At this pressure the boiling point of water is 100°C. However, as you rise above sea level, the atmospheric pressure becomes lower and water boils at a lower temperature. At about 3,000 metres above sea level, water boils at 90°C.

Ice is formed when pure water is cooled below 0°C. Water expands when it freezes and the molecules have more space between them, so ice is less dense than water. This is why ice floats on water. Floating bodies of ice such as icebergs are dangerous to shipping. The fact that ice does float is important to living things in the water because the top layer of ice stops the water underneath it from freezing. Other frozen forms of water are frost, hail and snow.

There are three states of water: 1 Solid (ice); 2 Liquid (water); 3 Gas (water vapour). At 0°C, pure water freezes and becomes ice. When heated to 100°C, water boils and forms a vapour called steam. This phenomenon occurs naturally in the form of hot springs and geysers. In its natural liquid state, water is both colourless and tasteless.

THE WATER CYCLE

The world always has fresh water because of what is called the *water cycle* or the *hydrologic cycle.* The two main forces which keep this cycle in progress are the sun's heat and the earth's gravity.

Evaporation and precipitation

Water heated by the sun is constantly being evaporated from the surfaces of rivers, lakes, seas and oceans. It turns to water vapour and mixes with the air and rises up into the atmosphere. As the air containing the water vapour **rises, it expands and cools. As the temperature falls, the air gradually loses the capacity to hold the water vapour.**

These threatening, dark clouds are rain clouds. Water evaporates from the surfaces of rivers, oceans, lakes and vegetation and rises to form clouds when it encounters colder levels of air. The cool air causes the water vapour to condense back into water droplets, which are visible as clouds. These droplets fall as rain, sleet or snow and eventually drain back into the seas and rivers to start the cycle all over again.

Finally, the air reaches a stage called the *dew point*, or saturation point. At this point, the air is saturated with water and further cooling causes the water vapour to condense to form tiny water droplets or ice crystals. These are the particles which form clouds. From the clouds, water droplets fall back to earth as hail, rainfall and snow. This is called *precipitation*.

A large amount of the precipitation that falls on the land is re-evaporated and goes through the water cycle all over again. Some is absorbed by the soil and is used by plants.

Runoff and ground water

Some of the moisture used by plants is returned to the atmosphere by transpiration, that is, excess moisture passes out through the plants' leaves, where it is evaporated. Some rainwater seeps through the soil into the rocks below as *ground water* and some flows across the surface and is called *runoff*. Runoff eventually makes its way into streams and rivers. All ground water and runoff that is not evaporated flows back to the sea to complete the water cycle there.

Around the North and South Poles and in high mountain ranges, precipitation mainly takes the form of snow. Some snow stays permanently at and above a certain height called the *snowline*. Here it may turn to ice, forming glaciers and ice sheets.

The water cycle can be shown diagrammatically. Water runs off the land and is collected in seas and lakes where the sun's heat causes it to evaporate. The vapour rises to form clouds, which are carried inland by the wind and deposit the water as rain over the land.

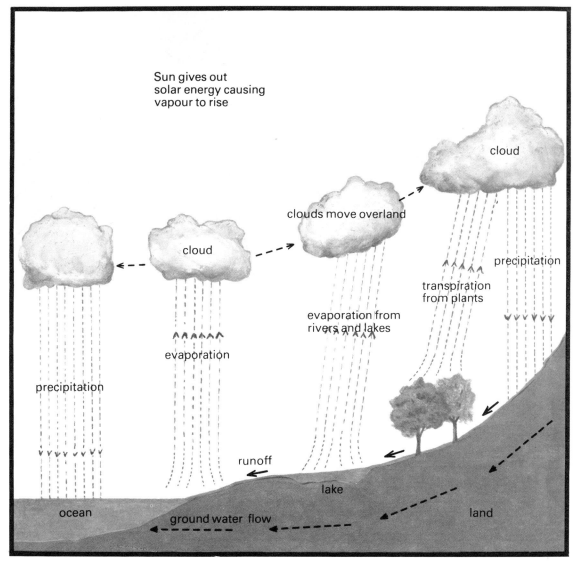

RIVERS, WATERFALLS AND DAMS

Rivers are large streams of water flowing downwards to the sea along channels they have cut. Some rivers flow into lakes and some join other rivers. When a smaller river flows into a larger one the smaller one is called a *tributary*.

Rivers

Rivers have played an important part in the development of civilisation. The valleys of such rivers as the Hwang Ho in China, the Nile in Africa and the Tigris Euphrates in south-west Asia were the homes of some of the earliest farming communities. Farmers in these valleys learned how to supply all their land with water drawn from the rivers by canals. Watering the land in this way is called

irrigation. Through irrigation they learned many other things including surveying and mathematics.

Rivers are one of the main ways in which rainwater returns to the sea in the water cycle. On their way to the sea, rivers help shape the land through *erosion*.

The development of rivers and their valleys is divided into three main stages. Rivers have a *youthful* stage, a *mature* stage and *old stage.* The youthful stage starts at the river source. Some rivers start as springs where clear water bubbles out of the ground. Some are the outlets of lakes.

Stages of rivers

In the *youthful stage,* the river's course is at its steepest. Although the volume of water in the river may be small, it flows quickly down the steep slopes. In wet weather, rain increases the volume of water and the small river may become a torrent. The fast-flowing water causes stones and gravel to rub against the rocks, smoothing them and

Rivers usually rise in hills or mountains where the rainfall is high. As they wind their way down the hills and across plains towards the sea, they gradually widen, gathering water from smaller tributary streams that flow into them. As the rivers cross plains, they often form meanders (bends) and oxbow lakes (cut-off bends) before reaching the sea.

Semi-circular oxbow lakes are formed when a river meanders across a plain and the outer bend of the meander becomes cut off from the main river. The river banks become eroded as new deposits of silt and other material build up as a result of flooding (1,2). Thus the river may change its course by cutting through the neck of the meander (3) as it becomes more restricted by deposits. Eventually the meander is completely cut off from the main course of the river (4) and forms an oxbow lake. In time, these lakes become swamps and gradually dry up.

deepening the river channel. This creates steep-sided V-shaped valleys.

In the *mature stage* the river has a larger volume of water but it flows down more gentle slopes. Mature rivers also cause erosion. They look muddy because they contain fine particles, *suspended load*, formed by the rocks in the water rubbing against each other and against the river bed. Some of the suspended load is soil carried from the sides of the channel. In mature valleys, rivers often flow in large bends called *meanders*. The river constantly washes away the outer banks of the meanders, widening the river valley, so that these valleys have much more gentle slopes than youthful valleys.

The *old stage* of a river valley occurs as the river flows over a nearly level plain. There is little erosion, but the river now carries along enormous amounts of suspended load that were eroded upstream.

In the old stage, rivers often overflow their channels when their volume increases. Because this stage is often flooded, the area is called a *flood plain.* Each flood spreads fine, fertile soil, or silt, over the land, so flood plains are among the best farmland in the world. During floods, silt is dumped along the river banks, forming mounds called *levees.* Mature rivers sometimes change their course and instead of flowing around a meander, they cut across the neck of it. When this happens water often remains in the meander, forming an *oxbow lake* which

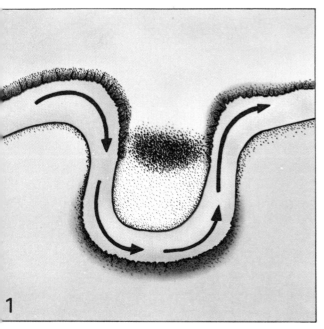

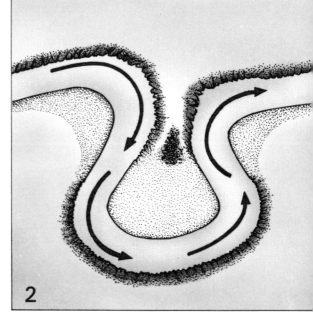

gradually becomes a swamp and may finally dry up.

If there were no interruptions to the work of rivers they would wear away the land until it was almost a flat plain. Sometimes geological movements of the earth lift up blocks of land over which rivers are flowing, or the sea level at their mouths may fall. Geologists say that such rivers are rejuvenated, or made young again.

Waterfalls and rapids

Many youthful rivers have very irregular courses, especially in areas where movement of the earth has uplifted sections of their courses. Common features of these rivers are *waterfalls,* formed when the river plunges suddenly over a steep slope or cliff. Waterfalls with only a small volume of water may be called *cascades* and those with a very large volume of water are sometimes called *cataracts. Rapids* occur on less steep slopes. The force of the water is sometimes used to generate electricity.

A waterfall can occur at places where gently sloping layers of hard rock lie on top of layers of softer rocks. The river erodes the softer rocks much more quickly than the hard rock and a ledge is cut out. Niagara Falls is placed where a layer of hard limestone overlies layers of soft shale and sandstones. The swirling waters at the base of the falls are constantly undercutting the softer rocks.

 New deposits

 Older deposits

 River channel and strongest currents

 River banks being eroded

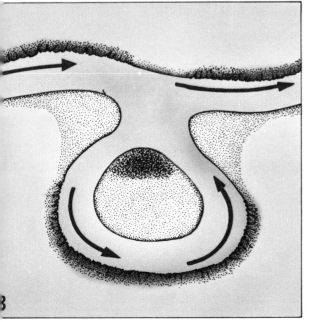

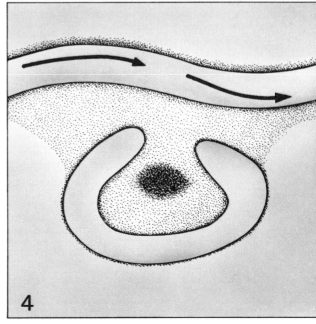

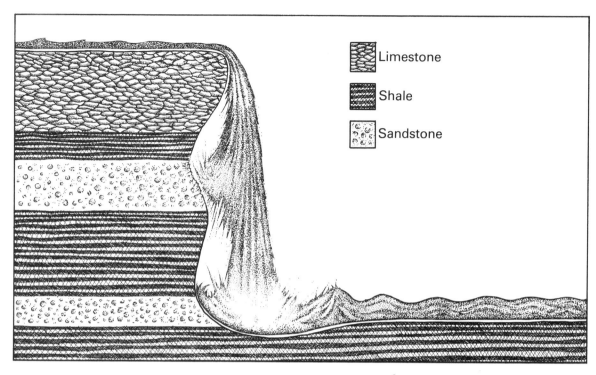

Waterfalls usually form when a ridge of hard rock blocks the course of a river. The force of the falling water gradually wears away the softer rocks below to undercut the top, hard layer of limestone. Great limestone rocks crash down into the swirling waters below as the falls retreat. From time to time, slabs of limestone break off and crash down into the water. Because of this, Niagara Falls is retreating back up the river at about one metre each year. All waterfalls are cut back in this way and the river course may ultimately be smoothed out.

Dams

Dams are barriers built to hold back flowing water. They are used to control floods, to store water for houses, factories and for irrigation and to generate hydroelectric power.

The design and method of construction of a dam depends on its purpose, where it is to be located and the materials and workforce that are available. Before a dam can be constructed, the water is diverted away from the site. This is often done by building tunnels upstream which carry the water around the dam site to a point downstream. *Concrete dams* can be of several kinds. The *gravity dam* uses its weight to keep it strong and stable. Every part of the structure is strong enough to withstand the water pressure on it. A cross-section of the dam looks like a right-angled triangle with the upright side against the water. *Arch dams* are less costly to construct. They are built against the sides of a valley and are curved towards the water in the shape of an arch. The shape of the dam gives it its strength; because it carries the water pressure to the sides of the valley, the wall does not have to be so thick.

The foundations for dams must be particularly secure. The site has to be excavated or dug out until solid rock is reached. Sometimes the rock may need extra strengthen-

The Eucumbene Dam in New South Wales, Australia, constructed as a part of an irrigation scheme, is an earth dam. The upstream face of most earth dams is lined with steel-reinforced concrete to prevent erosion and to increase the watertightness.

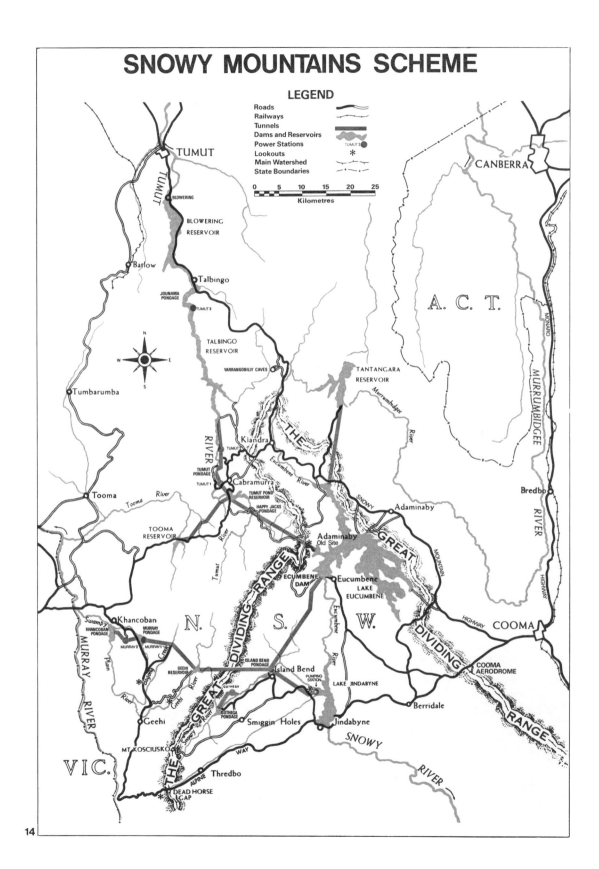

ing with steel and concrete. Gravity dams may be built of concrete alone (*mass* concrete), but in most other types the concrete is reinforced with steel.

Earth dams are usually made of rock and gravel as well as earth. The upstream face of the dam is lined with reinforced concrete or stone blocks to prevent erosion by the water and help to make the wall watertight. To build an earth dam, the material is usually dumped on the site and shifted into position with scrapers and bulldozers. The movement of heavy vehicles across the top of the wall as it is being built also helps to press down and compact the material beneath.

Weirs are low dams used to raise the level of the water in a stretch of river so that it may flow into canals for irrigation or for ships. Weirs are also used to control flooding, for several weirs on one river can hold back water to keep the overall flow more even.

Hydroelectricity

The power of water may be used to drive turbines. The source of the power may be natural, as in waterfalls, or it can be produced by damming rivers. A turbine is really a kind of rotary motor which is driven by water in a similar way to a water-wheel. In turn, it drives large generators to produce electrical energy. The first large hydroelectric scheme was completed at Niagara in 1892 and, since then, plants have been set up all over the world wherever there is sufficient water to do the work. The Snowy Mountains Scheme in Australia is an example, as is the

Opposite: The Snowy Mountains Scheme in Australia was one of the world's largest-ever engineering projects and took 25 years to complete. In order to produce hydroelectric power and provide extra water for irrigation purposes, the Eucumbene and Snowy rivers were diverted westwards under the Snowy Mountains through 145 km of tunnels, which were cut through the rock to transport the waters inland to the Murray and Murrumbidgee valleys.

Such dams as the Burrinjuck Dam (below left) and the Warragamba Dam (below), both in New South Wales, Australia, are impressive monuments of modern engineering. Dams may serve many purposes: storing water for irrigation or human consumption; controlling floods and erosion; diverting water into canals and pipelines; and raising the water to a sufficient level in order to produce hydroelectric power.

Dnieper Dam in the USSR and the Hoover Dam and Tennessee Valley Administration in the USA. In the South Island of New Zealand, one of the largest sources of power is the Roxburgh Dam on the River Clutha which feeds huge amounts of power into the South Island Grid System. A grid system is the network of power lines that take electricity from the places where it is produced to the points where it is used in homes and factories. The grid allows power from several power stations to be fed into a single network, making it possible to make up for interruptions to supply.

WATER SUPPLY

Many people in country areas get water from rivers, springs and wells. They also use rainwater collected in tanks and earth dams. Town and cities need much more water and use lakes, rivers and large *reservoirs*, or storage dams, often situated a long way from the cities themselves. Water from these sources is seldom pure. It usually contains minerals dissolved from the surrounding soil as well as particles of soil and decaying vegetation and, often, bacteria which can cause diseases. Such water is not always safe for drinking and must usually be treated with chemicals.

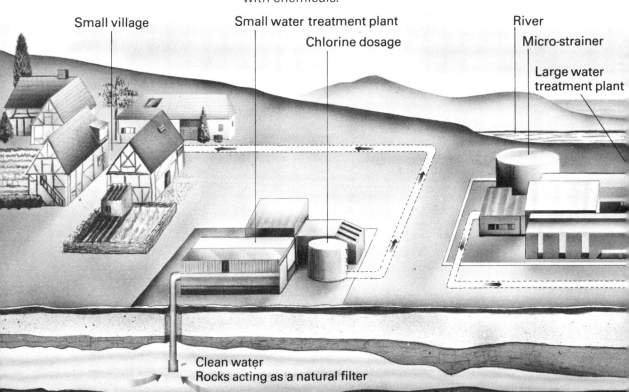

Purifying water

Water must be taken from the rivers or lakes at suitable places, called *abstraction points,* as far as possible from sewage works and factories which discharge waste products or *effluents.*

Water for a town or city is first piped to a water treatment plant. Here it goes through the first stage to make it pure, called *coagulation.* Coagulation is achieved by adding harmless chemicals such as alum to the water. These chemicals are known as coagulants because each of their molecules attracts and holds several particles of any impurities suspended in the water. When several particles join together they form larger, heavier particles which *precipitate*, or sink, in the water. Large clumps of such particles are called *flocs* or *aggregates.*

The water then passes into *settling tanks* where the heavy flocs, or aggregates, fall to the bottom and are removed. The coagulants, which are now in the centres of the flocs, are also removed in the process.

Coagulation removes much of the suspended matter, but some particles may not form flocs. The remaining particles may include bacteria so further treatment is needed. The second stage of treatment is often *sand filtration*. The water is pumped to a large concrete

Water is collected and stored in reservoirs and is then distributed according to demand from consumers. A pumping station pumps water out of the reservoir and it is passed through underground pipes to the water treatment plant, where it undergoes a purification process to dispel unwanted industrial pollutants and effluents. It is treated with chlorine and strained and filtered to remove toxic substances. The water is then channelled to both industrial and domestic consumers through a network of pipes. In rural areas there are smaller treatment plants to cater for local demand.

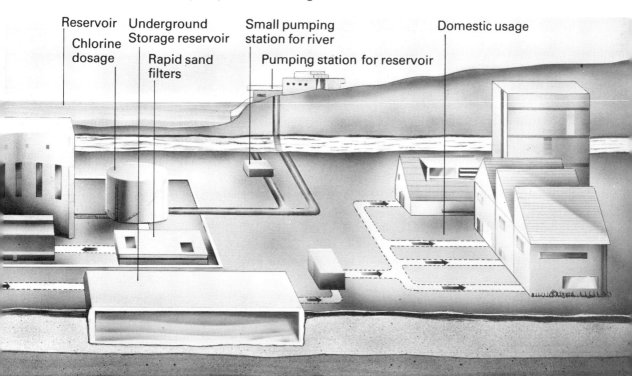

Before the water that we use and drink reaches the taps in our homes, it undergoes a lengthy process of purification. The first stage in water treatment involves coarse screening to remove leaves, debris and other unwanted material. When the water enters the water treatment plant, it is given a dose of chlorine to control bacteria. Then it is passed through a micro-strainer, which traps minute particles and algae. The water is channelled through flocculation and sedimentation tanks before passing through sand filters. The clean water may be treated in an aerator where it is saturated with oxygen. It is given a chlorine dosage and passed by powerful pumps to the mains supply for distribution.

chamber at the bottom of which is a bed of fine sand several feet thick. The water filters through this sand and much of the remaining suspended material is removed.

However, because bacteria may be extremely small, a third stage of treatment is necessary. This is called *chlorination.* Here, chlorine gas is bubbled through the water and a small amount is dissolved. At the same time, the water is thoroughly mixed so that it is all treated. Chlorine kills all disease-causing bacteria that are likely to be in the water. Some bacteria form spores which are not killed by chlorine, but these bacteria do not cause disease when present in small numbers in drinking water.

Water softening

Natural water usually contains a certain quantity of minerals like calcium or magnesium. Calcium and magnesium salts react with soap to form a scum instead of a foaming lather. When these things happen we say the water is hard and the water must be softened by a special process to make it more pleasant for household use. Sometimes this is done at a water treatment plant, especially where factories must use **soft water,** but often hard water is treated at home in the household water supply.

Water softening removes calcium and magnesium from the water. Treatment with lime and soda reduces the amount of dissolved salts in water. The lime removes the calcium and the magnesium is removed by the lime and soda.

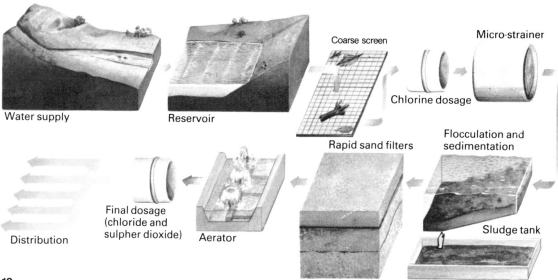

Water supply
Reservoir
Coarse screen
Chlorine dosage
Micro-strainer
Rapid sand filters
Flocculation and sedimentation
Sludge tank
Aerator
Final dosage (chloride and sulpher dioxide)
Distribution

De-ionization is the name of the process which removes all salts from water, including those that cause hardness, to produce very pure water. To do this, the water is passed through a tank filled with *zeolite,* a substance which contains small holes filled with sodium ions (tiny electrically charged particles). These sodium ions are easily replaced. When the water flows through the zeolite, it dissolves the sodium ions and leaves calcium and magnesium ions in their place. Flushing the tank with a solution of salt returns the sodium and removes the calcium and magnesium as compounds which are soluble in water and can be rinsed away.

After water treatment, the purified water is usually pumped to underground reservoirs or storage containers beneath cities and towns. From these reservoirs, the water is pumped to homes, offices and factories.

Country dwellers who have no supply of treated water sometimes have to boil their water before drinking it. This kills some of the bacteria. Water may also be purified by boiling it and collecting and condensing the steam, a process known as *distillation.*

Top: In the secondary sedimentation tanks of the water treatment plant, the water is filtered to remove any impurities.
Above: Water is passed through an aerator after it has been filtered. In the aerator, it is saturated with oxygen before being sterilized with chlorine.

THE OCEANS

Oceanography

The study of the oceans is called *oceanography*. Oceanographers study four main things about the oceans: their physical features (water temperature, the movement of ocean waters and behaviour of ice in the sea); their geological features (the study of the ocean floor and how features in the deep waters are formed); biological features, also called marine biology (the study of life in the sea); and the chemical composition of the sea water.

Before the mid 1800s, little was known about what lay under the sea. It was not until the first submarine cables were laid in the 1850s that people started to be interested in the ocean floor.

The echo sounder

In 1919 the echo sounder was invented. This instrument helps oceanographers to make maps of the bottom of the sea. It works by sending sound waves down from the ship to the ocean floor. The echo that bounces off is picked up

Land constitutes only a small percentage (29 per cent) of the earth's surface - water covers the rest. The oceans are a valuable source of chemicals, food and minerals, in addition to acting as a huge reservoir for solar heat energy.

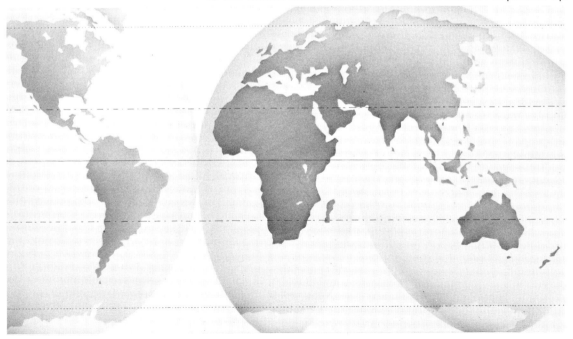

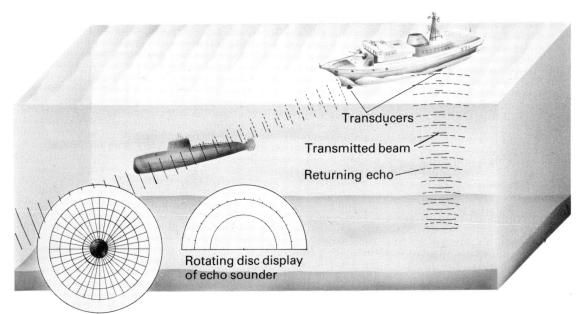

Cathode-ray display of Submarine detector

by a very sensitive receiver. Attached to the receiver is a device which works out how long the echo takes to return. The deeper the water, the longer this will be, so by using this time and knowledge of how fast sound travels in water, it is possible to calculate underwater distances accurately.

In the 1870s, a British ship, the HMS *Challenger,* took a measurement of depth, called a *sounding,* of part of the Pacific Ocean by lowering a heavily weighted line down to the bottom. The depth turned out to be 2,425 fathoms (a fathom is equal to six feet or about 1.85 m) and it took the sailors two and a half hours to let down the rope and haul it back. Using the echo sounder, it would have taken only a few seconds to work out the depth of the water.

Submarines, shoals of fish and wrecks on the sea bed can all be detected by modern sound-echoing devices. An anti-submarine craft has two transducers, which convert electrical power into sound. One transducer transmits two signals to measure the water depth and these are displayed on the rotating disc of the echo sounder. When a submarine is detected, it appears as a dot on the cathode-ray display.

Diving machines

Oceanographers use instruments to take samples and measure water temperature at various depths. They also collect samples of mud and rock from the sea bed to study and have experimented with ways for man to go under the sea to explore it for himself. One of the earliest such machines was the *diving bell,* which was developed a long time ago to enable divers to build bridge foundations underwater. One of the first underwater vessels was a *bathysphere,* a hollow metal ball lowered from a ship on a cable. Inside the diver had air and light and a telephone line to talk to people on the ship, but he could not move about very much.

An improvement on this was a *bathyscaphe* (from two Greek words meaning 'deep boat'). This vessel could move up and down under its own power and also move a little on the ocean floor, so divers could explore much more under the sea.

More recently, very small submarines for one or more divers have been able to descend to great depths. Divers use these as a base where they can rest and eat, then go out to explore the surrounding waters in special diving suits with tanks of oxygen strapped on their backs. The diving suits used at these great depths are pressure-resistant. The word *scuba*, an acronym of '*s*elf-*c*ontained *u*nderwater *b*reathing *a*pparatus' is usually used to describe diving equipment. Because of the effects of water pressure, normal scuba diving is possible only in shallow water.

The tides

In most parts of the world the sea level rises and falls about twice a day. These are the tides. Tides are caused by

Below: Divers often use a diving bell when working underwater. This is the Arms Diving Bell, which is depth rated to 1,000m. However, modern bathyscaphes can descend to depths of 12,000m.

Below right: Diving suits are now very advanced and can withstand pressures of over 300m. This atmospheric diving suit is depth rated to about 700m. The diver breathes air at normal atmospheric pressure from the oxygen back-pack. He can work with tools by means of the lever-controlled fingers that reach out of the large round gloves.

the strong gravitational pull, or force, of the moon on the waters of the earth. The sun also exerts a gravitational pull, but it is much smaller.

Though the pull of the moon acts most strongly on the water directly below it, the effect sometimes occurs before or after the moon passes a particular point, for the rotation of the earth and the position of the sun, as well as the earth's own gravity, all increase or decrease the effect of the moon's attraction.

In many places there are two high tides and two low tides in each twenty-four hours, while in others there may be only one high and one low tide per day. The reasons for the variations in *tidal range,* or difference between high and low tides, around the globe are complicated. In some landlocked seas like the Caspian and the Mediterranean, the tidal range may be measured in a few centimetres, while in other places it is metres. The time of day at which high and low tide is reached varies too.

This diver is working underwater on platform inspection. His suit is made of magnesium alloy with jointed arms and legpieces. Many divers are now engaged in often-dangerous work on offshore oil platforms. Fatalities and injuries are not uncommon and the 'bends' are an ever-present hazard of the occupation if a diver ascends to the surface too rapidly. He therefore ascends in several stages that are specially prescribed in decompression tables.

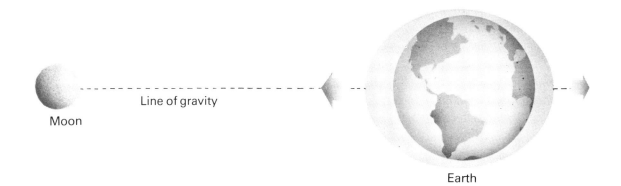

The surface of the earth is encircled by two high-water periods, one of which is the result of the gravitational attraction of the moon. The sun and the moon both tend to pull the water in the seas towards them and thus as the earth rotates on its axis the water levels change. To balance the rise in the water level on one side of the world there is a high tide on the opposite side. Low tide is midway between the two.

Tides tend to have the greatest range in narrow passages, or straits, such as in the Bay of Fundy, where the rush of water between the shores forces it to build up.

There are two high tides each day in most places because there is always a high tide on the opposite side of the earth to the moon as well as one on the side facing the moon. This is because of two main gravitational forces acting on the water. The moon exerts a pull away from the earth thus causing high tide on the side of the earth facing the moon. Another force acting on the water is the pull of the earth's own gravity. However, the high tide on the opposite side of the earth to the moon is caused by *centrifugal force* since at that point, although the pull of the moon in this position combined with that of the earth would normally draw the water towards the centre of the earth, the spin of the earth counteracts this sufficiently to make the water build up into a high tide.

On the shore, you can see the high tide coming in as the water slowly moves up sloping beaches and shores until there is often just a small stretch of beach left showing. When the tide goes out, all of this land is uncovered again. The level of the water also changes in the middle of the ocean, but the differences are small and very hard to measure, even with special instruments.

The highest or *spring tides* happen when the sun, moon and earth are in a straight line. This means there is a combined gravitational pull acting to make the high tides occur twice each month when the moon is full and when there is a new moon.

The lowest or *neap tides* occur when the sun, earth and moon form a right angle. The pull of the sun and moon act against each other to make the high tides lower and the low tides higher than usual so you cannot see so much difference between them. Neap tides occur when the moon is in its first and third quarters.

Tidal power

Where there is a great difference between the high and low tides, the movement of the water can be used as a source of electric power. A dam or barrier may be built across an estuary or tidal mouth of a river. Sliding gates or sluices are opened in the barrier to allow the tide in, then closed at high tide. When the tide outside has fallen again, water inside the barrier is released through turbines to generate electric power.

Currents

Currents are movements of water on and under the surface of rivers, seas and oceans. Sometimes they have definite boundaries and are like rivers flowing through the ocean. These are called *streams.* Currents which do not have such boundaries are called *drifts.* How warm or cold the water is and how salty it is will affect how the current moves through the rest of the ocean.

Scientists have found that in the northern half of the world, the *northern hemisphere,* ocean waters move in a

Currents are great, continuous flows of water in the oceans of the world. They are now being studied by oceanographers who immerse special current meters in the sea to record information about the nature and characteristics of currents.

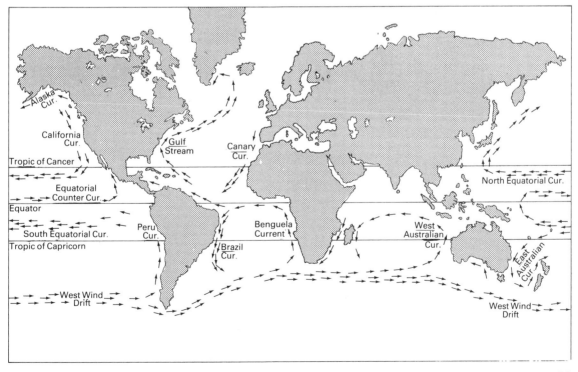

clockwise direction, while in the *southern hemisphere,* the waters move in an anti-clockwise direction. Under the surface, some currents flow in a different direction to the main body of water. Other water movements like *upwelling,* the vertical movement of water, bring minerals and other material from the bottom to the surface and tiny sea life such as plankton can then feed on it.

The ocean currents stream around the edges of the continents in five great circuits around the north and south Pacific, the north and south Atlantic, and the Indian Ocean. Among the well known currents is the great Gulf Stream that runs from the coasts of Central America north-west past the coast of Ireland. Its warm waters make the climate of much of the northern coast of Europe a great deal warmer than it would otherwise be.

These surface currents move quite slowly, mostly not over 1 km/h, though in exceptional cases speeds of 4, 5, 9 and even over 17 km/h have been measured.

Most of the water in the oceans is moving slowly but constantly. Not only are there surface currents, there are also slow deep-sea currents, such as those that bear the cold deep waters towards the Equator. Besides the upwelling movements of rising water there are also several places where cold ocean waters sink to the bottom. Three such places are in the south Pacific, the south Atlantic and in the north Atlantic between Iceland, Spitzbergen and Greenland.

Ocean currents circulate the waters, distributing oxygen and the sun's heat and food of all kinds for the teeming life of the seas.

Water waves are transverse waves, the vibrations being at right angles to the direction of the wave. Thus a boat moves up and down as waves pass horizontally. Actually, the water particles rotate, each one describing a short, vertical, circular course.

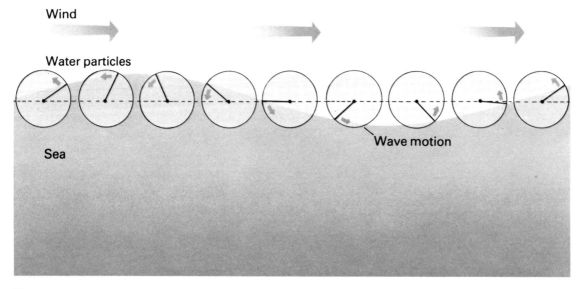

Waves

Waves seem to be masses of water that travel over great distances, but the wave travels through the *water,* which does not itself move a great distance, though the *waves* may come from thousands of kilometres away.

The movement of the water in a wave is up and down, as though in a circle, rather than forward. Out at sea, the water is moved up and down as the great waves travel through it, close to shore the movement is slowed by beaches or rocks and the wave 'breaks'.

Ocean waves seem to move water around, but they actually move the water backward and forward along the coastline, for the shore has a slowing down effect, or *drag,* on the water. Out in the open sea, the water molecules move up and down. This is why a cork will bob up and down in the sea without changing its position, unless it is blown by the wind.

The ocean bed

The land under the sea is as varied as it is on the shore, with mountain ranges and valleys and deep gorges called *ocean trenches.* When a mountain rises above the water it is known as an island. Where shallow seas cover part of the land, this is known as the *continental shelf.* As you travel out from the shore, the continental shelf ends suddenly at a steep slope called the *continental slope* which runs down to the deepest parts of the ocean, called *abysses.*

Where an ocean floods the edges of a land mass, a continental shelf is formed near the shoreline. This shelf ends in a slope that gradually runs down towards an abyss, which is made up of hills and flat plains, sometimes intersected by deep ocean trenches. Around the long underwater mountain chains (mid-oceanic ridges) lie rocks that were formed comparatively recently in the history of the earth.

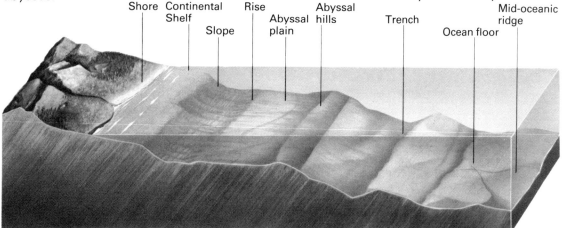

Ocean resources

The basic food in the ocean is *plankton,* which is made up of millions of tiny plants and animals. Other animals including fish live on plankton. The smaller fish are eaten by the larger fish, which in turn may be eaten by man. In some parts of the sea, too many fish have been caught and some sea creatures are dying out or becoming extinct for lack of food.

To prevent this, oceanographers are looking for ways of farming the sea. Just as we create the best conditions for animals to survive and grow on land, *aquaculture* tries to create the best conditions in the sea for fish to thrive.

The oceans contain many minerals. Between 3.3 and 3.7 per cent of sea water is salt. Sea water also contains huge amounts of many minerals such as gold, magnesium and potassium. Taking most of these minerals from the sea is too costly at present, while we still have supplies on land, but scientists are investigating ways to make mining the sea more efficient and less expensive in the future.

The German stern ramp trawler drags its nets through the sea behind it. A more efficient method of trawling than the side-method used by the standard trawler (inset), it enables the working wires to run free of the deck, and larger catches of fish are hauled aboard. Many modern trawlers have refrigeration units so that freshly caught fish can be frozen immediately.

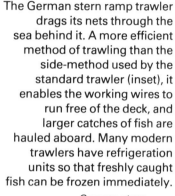

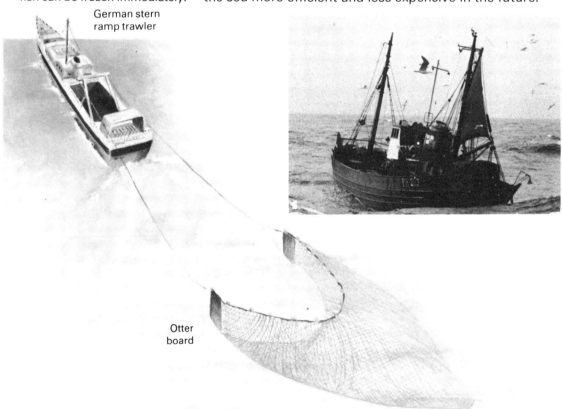

German stern ramp trawler

Otter board

HARBOURS, DOCKS AND NAVIGATION

Harbours are sheltered stretches of water that provide safe places for ships to anchor. A *port* is a harbour with quays, landing places, and facilities for loading and unloading ships.

Some of the finest harbours in the world are natural ones. Sydney Harbour in Australia and Rio de Janeiro in Brazil are outstanding examples. But most harbours are artificial ones, formed by building a protecting wall called a *breakwater* or *mole* out into the sea to enclose a stretch of water. The breakwater must be very strong to withstand the great forces caused by winds and waves.

Sydney Harbour, Australia, is possibly the most beautiful natural harbour in the world, providing a safe refuge for ships. Surrounded by land, it has access to the sea through narrow channels only. The modern city of Sydney has sprung up around the harbour and port. Notable landmarks are the Harbour Bridge and Opera House.

Breakwaters

Two main types of breakwater may be constructed. The blockwork type is made of an upright wall of stones or concrete blocks each weighing fifty tonnes or more. The rubble-mound type is a long mound of stone. Often, the two types of construction are used together, with stone rubble providing the foundation for a masonry superstructure.

Ocean and river currents are continually depositing silt on the floor of a harbour. *Dredging* is the process used to remove this mud and other material so that navigation channels are kept free and there is always enough depth of water for ships. When a new harbour is designed, local conditions of winds and tides have to be considered and

The stationary bucket dredger, still in use today, is usually towed by tugs to the site that is to be dredged. It is anchored to the sea bed and a bucket chain conveyor scoops up mud and silt from the water below. The frame can be hydraulically raised or lowered to the required level. The bottom tumbler shaft must always be totally immersed to avoid any overheating.

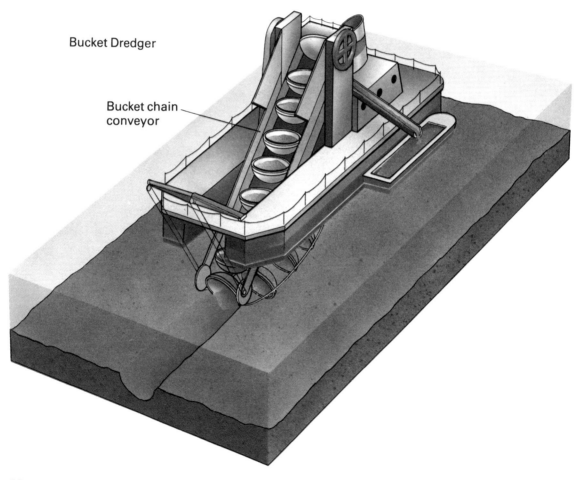

Bucket Dredger

Bucket chain conveyor

so scale models are built to test how the design will work.

Within the protection of the harbour ships can unload their cargo, but the tides have to be considered. The tidal range, or amount of rise and fall of the tides, varies a great deal around the world. Such seas as the Gulf of Mexico, the Baltic and the Mediterranean have very weak tides. But other places have an enormous tidal range. Off the port of Broome in Western Australia the tidal range is great. Whenever the tidal range is more than about three or four metres, it is necessary to provide *docks* for loading and unloading.

Docks

These docks are enclosed areas into which the ship can sail at high tide. Lock gates are closed behind the ship so that, when the tide falls, the water level inside is not affected. The lock gates may be made of timber or braced steel, but timber lasts longer because the sea water corrodes or eats away the steel. The dock entrance may also be closed by hollow steel structures which are slid or floated into position. These are known as *caissons.*

Graving docks are used for ship repairs. They are enclosed basins with walls made of concrete or masonry.

These pearling luggers are beached on their sides at low tide on a beach in Western Australia. When the tide flows in, they will refloat and right themselves. Ships are often beached while the tide is out in areas where the tidal range is great.

In this floating dry dock, the A.N.L. vessel *Empress of Australia* is undergoing an inspection and repairs. When the water is pumped out of the dock, the submerged dock is raised by means of ballast tanks. Huge docks, such as this, can lift heavy ships of up to 70,000 tonnes deadweight.

Because they can be emptied of water, they are also known as *dry docks*. They allow workmen to reach parts of the ship's hull which are normally underwater, so the ship can be repaired or painted, or scraped clean of barnacles and shellfish which cling to ships' bottoms.

Many ports are also equipped with *floating docks*. These are open-ended structures made of hollow steel in the shape of an L or U. They have chambers which can be filled with water, then emptied. When it is filled with water, the floating dock sinks down far enough to let the ship sail into it. The ship is supported on a cradle of heavy beams and the water is pumped out of the dock chambers. The dock rises up on top of the water, carrying the ship up with it.

Floating docks are useful because they can be taken anywhere they are needed, from port to port. They can accommodate ships longer than their own length, because of their open ends. Floating docks which are L-shaped can also accommodate ships wider than their own width. If the ship to be repaired cannot float upright in the water, the floating dock can be trimmed or tilted so that it can hold the ship.

In the city, street signs tell you where you are. In the country, you can note the position of hills, rivers and any other features or landmarks. On the sea, especially if you are out of sight of any land, you must use other means to find where you are and work out the direction you wish to travel. Doing this is called *navigation.*

Early navigation

Early sea navigators sailed without any instruments. Whenever possible, they kept within sight of land, seldom daring to venture out into the open sea.

One of the first important navigational instruments was the *astrolabe*, which was invented around 150 BC. This instrument was used to measure the angle of stars and other heavenly bodies above the horizon. Using this instrument and keeping records of direction and distance helped early navigators work out very roughly where they were.

Soon after 1200 AD, European sailors learned how to use a simple magnetic compass, which shows which direction is north. The principle of the compass had been known in China for at least 200 years before. With the

This diagram shows the front and back views of an astrolabe, used by astronomers and by navigators to determine their position at sea. The general principle of the astrolabe is that it uses two beams of light from the same star, one of which is direct and the other reflected from a mercury surface, and these are directed into a telescope. When these two beams become parallel, the images coincide, and the star is then at the standard altitude. The principal parts of the astronomer's astrolabe are: 1 Ecliptic ring; 2 Index; 3 Star pointer; 4 Tropic of Capricorn ring; 5 Zenith; 6 Horizon; 7 Unequal hour line; 8 Equal hour scales; 9 Alidade; 10 Sight; 11 Scale of degrees; 12 Zodiac calender scales; 13 Shadow square.

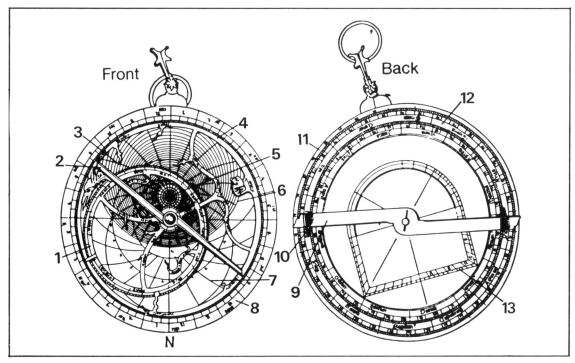

Modern navigational instruments are much larger and more complex than those of the past. This helmsman on a large car ferry does not use a traditional ship's wheel for steering. The modern steering console in front of him has a switchboard fitted with elaborate dials and switches, which he uses to steer the ferry.

compass, navigators could set a course in the direction they wished to go.

A common method of navigation was *dead reckoning*, which was the way Columbus navigated his ship when he discovered America in 1492. He set his course with a compass and marked his chart or map to show the distance he thought he had sailed along a straight line. He was able to guess the speed of the ship by watching its movement through the water. He measured time with a half-hour sand glass.

Navigators also measured speed by throwing overboard a log attached to a line. This line had knots tied in it at regular intervals. As the log floated away, a sailor counted the number of knots against time measured with a sand glass. This is how *knots* rather than miles per hour came to be used to describe the speed of ships.

Modern navigation

There are four main ways of finding your way at sea: *piloting* in coastal waters, *dead reckoning* and *celestial navigation* in the open sea, and modern *electronic navigation*.

Pilots are seamen who have special knowledge of local currents and tides. They use exact charts which show the depth of the water and the heights and positions of various landmarks. They also use a weighted line (which was once called a *lead*) to measure the depth of water around the ship and compare it with their charts and they also make accurate observations of the positions of landmarks on shore.

Celestial navigation means finding a ship's *latitude* and *longitude* by measuring the positions in the sky of heavenly bodies including the stars and the sun.

In modern times, electronics has greatly changed the nature of navigation. Now, seamen use the echo sounder to measure the depth of water simply and accurately and radio-direction finders which show where the ship is in relation to several transmitting centres. Radar helps navigators to work out the distance and direction of any point including coastlines or floating icebergs and the magnetic compass is often replaced by a **gyrocompass**. Some artificial satellites even transmit radio signals which ships can receive to allow them to fix their positions.

Robert Hooke's chronometer was designed to keep time accurately at sea and became an essential navigational instrument. Hooke perfected his design after studying watch mechanisms. However, the first chronometer was invented in 1714 by John Harrison, a Yorkshire carpenter, who won the £20,000 prize offered by the British government for such an invention. This original design was later modified and improved by John Arnold and Thomas Earnshaw and, later, Hooke.

Latitude and longitude

Latitude and longitude are measures that are essential to navigation. With a very accurate clock or watch called a *chronometer*, a way of estimating the ship's speed, the *log*, an instrument called a *sextant* which was developed from the old astrolabe, a *compass* and the charts of earlier navigators, Captain James Cook was able to navigate accurately enough to explore vast areas of the Pacific and Arctic and Antarctic. But he needed all those instruments to measure latitude and longitude correctly. Globes used by geographers and navigators are marked in *degrees of latitude* to show distance north or south of the Equator, and *degrees of longitude*, which show the distance to east or west of a line passing through Greenwich in England. This is the position agreed upon as 0° of longitude, just as the Equator is taken as 0° of latitude. Thus these two

The navigational sextant is an invaluable instrument for sailors wishing to pinpoint their position at sea. The seaman holds the sextant in his hand and looks through the eyepiece (F). He adjusts the position of the mirrors until the image of the sun appears to touch the horizon. The sextant consists of a light brass framework (ABC) and a vernier (at D), which can be read through a microscope. When the observer looks through the telescope (F), he can see the horizon (H) through the unsilvered half of the mirror (E). The light from the sun or a star (S) is reflected from the index glass (C) to the silvered part of the mirror into the eyepiece. The arm (CD) is used until the images of the star and the horizon coincide.

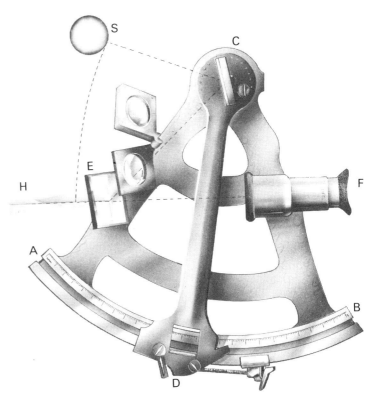

systems of imaginary lines on the globe give navigators a *grid* to which their charts and maps, and therefore the actual position of their ships, can be related.

WATER POLLUTION

Opposite: Public concern is increasing about the gradual pollution of our environment, especially of the rivers and seas, and many pressure groups have sprung up to call for government action. Many rivers are polluted and emptied of fish and aquatic life after the indiscriminate dumping of poisonous industrial waste. Most of this toxic waste eventually flows into the sea where it is usually absorbed.

One of the biggest pollution problems concerns water. Although we know how important water is to all living things, we are only just now beginning to think about how we can look after the earth's water for ourselves and for future people.

Waste products

Almost everything man does causes pollution of some kind. Household sewage, our own waste products, must pass through special sewage treatment plants to kill the germs which carry disease. Unless the sewage is treated before it is discharged into rivers, lakes or seas it will soon

Water pollution may take many forms. Rivers and seas may be polluted by sewage (below), and this will appear as a blue area in an infra-red photograph. Sometimes, oil (below right) and foaming household detergents (bottom) are discharged into streams and rivers. Trippers and holidaymakers often pollute beaches and riverbanks with their rubbish (bottom right).

kill the life of the waterways and contaminate the fresh water we need.

Many things in nature naturally break down into harmless substances. When this happens, we say they are *biodegradable.* Biodegradable products will rot away eventually and break down into substances that can be used by living organisms. This happens when compost is made out of vegetable peelings and waste. The vegetable matter breaks down into a substance which can be used to feed new, growing plants.

However, some industrial waste products are not broken down. They make the environment dirty, clutter the landscape and make the waterways smell. They can even be a serious threat to our health. For example, a

plastic bag that is thrown away stays where it is dropped and does not break down, unlike paper or timber which will eventually rot away.

Poisoning the seas

Only in the last couple of decades have we started to realise that man cannot go on dumping his rubbish into the sea forever. Waste products from the land usually find their way into the sea sooner or later. Then the poisons are absorbed by the tiny floating sea creatures called plankton which feed larger creatures, so the larger creatures are poisoned and this goes on up the 'food chain' until the largest creatures, the fish, are also poisoned. In Japan, people have already been poisoned by eating fish caught in waters polluted with waste industrial chemicals.

Many pollutants are finding their way into the sea and damaging the food chains. Pollutants include the sprays used to control insects on farm crops and the chemical fertilizers used to grow them. Fertilizers contain chemicals such as nitrates and phosphates, which make algae and other water plants grow so fast that the waterways become choked with vegetation. The plants use up the oxygen in the water and soon the animal life starts to disappear. Many detergents which are not biodegradable

Top: Oil can be dispersed from holiday beaches by spraying it with detergents, as shown here after the oil spillage from the *Torrey Canyon* oil tanker affected Cornish beaches. Above: Birds are often the innocent victims of pollution from fertilizers and chemicals.

Left: many canals and rivers are polluted by rubbish and detergents. Some of these polluted backwaters are now being cleared by concerned environmentalists and are then restocked with fish.

also contain phosphates which cause problems when they find their way into the sea.

Oil leaking or dumped from ships is a main pollutant of the world's waters. The oil does great harm to sea life and leaves a thick black layer on beaches. Oil tankers and offshore oil wells spill a great deal of oil into the seas.

Many countries impose very high fines on owners of ships that spill oil into the ocean. In this way, people are made to think more about the environment and be more careful.

Trying to stop pollution

As scientists, politicians and other people become more aware of the dangers of pollution, they are working to stop pollutants finding their way into the environment.

In many countries new laws have been passed to control the pollution of water and air. Cars and other machines are now being designed so that they do not cause so much pollution from exhaust fumes.

Scientists and agriculturalists are also finding new ways to reduce the amount of synthetic fertilizers they use on farmland and to use less harmful sprays to control insects on crops. Sometimes, pollution is just a nuisance, but poisoned water, smog and waste oil are not only unpleasant, they are a danger to the health and the life of the world.

Smoke and swirling, yellow smog are not an uncommon sight in the sky over many industrial towns. However, new clean-air controls and restrictions are being introduced by governments to limit this serious form of pollution. Many cities have been designated as smokeless zones and only specially manufactured smokeless fuels can be burnt in homes.

INDEX

WATER 1-5
THE WATER CYCLE 5-7
RIVERS, WATERFALLS AND DAMS 8-16
WATER SUPPLY 16-19
THE OCEANS 20-28
HARBOURS, DOCKS AND NAVIGATION 29-36
WATER POLLUTION 36-40

Page numbers in italics refer to a diagram on that page.
Bold type refers to a heading or sub-heading.

A

Abstraction points 17
Abysses 27, *27*
Aerator *18, 19*
Aggregates 17
Alluvium 2
Alum 17
Aquaculture 28
Arms Diving Bell *22*
Astrolabe 33, *33*
Atmospheric pressure 4-5

B

Bathyscaphe 22
Bathysphere 21-22
Bends 23
Bay of Fundy 24
Bicarbonate of calcium 4
Bicarbonate of magnesium 4
Biodegradable 39-40
Blood 1
Boiling point (water) 5
Breakwaters 29, **30**
Burrinjuck Dam *15*

C

Caissons 31
Calcium 18-19
Cascades 11
Caspian Sea 23
Cataracts 11
Cathode-ray display *21*
Centrifugal force 24, *24*
Challenger, HMS 21
Chlorination 18, *18*
Chronometer 35, *35*
Clouds 6, *6*
Coagulation 17
Columbus 34
Compass, magnetic 33-35
Continental shelf 27, *27*
Continental slope 27
Cook, James 35
Currents **25**, *25*, 26

D

Dams **12**, 12-15, *13, 14, 15*
  arch 12-13
  concrete 12
  earth *13*, 15
  gravity 12-15

Dead reckoning 34-35
De-ionization 19
Deltas 2
Detergents *38*, 39-40, *39*
Dew point 6
Distillation 19
Diving bell 21
Diving machines **21**, 22, *22*
Diving suits 22, *22, 23*
Dnieper Dam 16
Docks **31**, 31-32
Dredging 30, *30*
Drifts 25
Dry docks 32, *32*

E

Earth's surface *20*
Echo sounder **20**, 21, *21*, 35
Effluents 17
Electrolysis 4, *4*
Empress of Australia 32
Erosion 9-11
Eucumbene Dam *13*, 14
Evaporation **5**, 6

F

Fertilizers 39-40
Floating docks 32, *32*
Flocculation tanks *18*
Flocs 17
Flood plains 2, 10

G

Glaciers 2, *3*, 7
Graving docks 31-32
Greenwich, England 35
Grid 36
Ground water 6, 7, *7*
Gulf Stream 26
Gyrocompass 35

H

Harbours 29
Hard water 4, 18-19
Harrison, John 35
Hooke, Robert 35
Hoover Dam 16
Hwang Ho River 8

Hydroelectricity **15**, 16
Hydrogen 2-4, *4*
Hydrologic cycle 5-6, *7*

I

Ice 4-5, *5*
  sheets 2, 7
Icebergs 5
Irrigation 8-9

K

Knots 34

L

Latitude **35**, 36
Lead 35
Levees 10
Life **1**
Log 35
Longitude **35**, 36

M

Magnesium 18-19
Meanders 9, 10, *10-11*
Mediterranean Sea 23
Micro-strainer *18*
Mid-oceanic ridges 27
Mole 29

N

Navigation
  celestial 35
  early **33**, 33-35
  instruments *34*
  electronic 35
  modern **35**
  piloting 35
Neap tides 24
Niagara Falls 11-12, 15
Nile River 8

O

Ocean bed **27**, *27*
Oceanography **20**
Ocean resources **28**
Ocean trenches 27
Oil pollution *39*, 40
Oxbow lakes 9, 10, *10-11*

Oxygen 2-4, *4*

P

Pearling luggers *31*
Perspiration *4*
Photosynthesis *1*
Plankton 26, 28
Plants 1-2, *1*
Poisoning seas **39**, 40
Pollution 36-40, *37, 38, 39, 40*
Ports 29
Precipitation **5**, 6
Purification (water) **17**, *16-17*, 18, *18*

R

Radar 35
Radio-direction finders 35
Rain clouds *6*
Rainwater 7, *7*
Rapids **11**, 12
Reservoirs 16, *16-17*
Rivers *2*, **8**, 8-11, *9-10, 10-11*
  stages **9**, 9-11
Roxburgh Dam 16
Runoff **6,** 7, *7*

S

Sand filtration 17-18
Satellites 35
Saturation point 6
Scuba diving 22
Sedimentation tanks *18*, 19
Settling tanks 17
Sewage 36-39, *38*
Sextant 35, *36*
Silt 2
Smog *40*
Smokeless zones *40*
Snow 7
Snowline 7
Snowy Mountains Scheme *14*, 15
Softening (water) **18,** 19
Soft water 18-19
Solvents 4
Sounding 21
Spring tides 24
Streams 25
Submarine cables 20
Suspended load 10
Sydney Harbour 29, *29*

T

Tennessee Valley Administration 16
Tidal power **25**
Tidal range 23, *31*
Tides **22,** 22-24, *24*, 30-31
Tigris Euphrates River 8
Torrey Canyon 39
Transpiration 6-7
Trawlers *28*
Tributary 7
Turbines 15

U

Upwelling 25

V

Vapour pressure 4-5

W

Warragamba Dam *15*
Waste products **36,** 36-39
Water **1**, 2, *4, 5*
  cycle 5-6, *7*
  formula 2-4, *4*
  hard 4, 18-19
  physical states 4, *5*
  pollution 36-40, *38, 39*
  properties **2**, 2-5
  purifying **17**, *17*, 18, *18*
  soft 18-19
  synthesis 4
  underground 2
  vapour 4, *5, 6*
  weathering 2, *2*
Waterfalls **11**, 12, *12*
Waves *26*, **27**
Weirs 15

Z

Zeolite 19